snow monkeys

snow monkeys

PHOTOGRAPHS BY MITSUAKI IWAGO

TEXT BY HIDEKO IWAGO

CHRONICLE BOOKS

SAN FRANCISCO

pages 4,5: In 1970, Japanese monkeys from Jigokudani, or Hell Valley, in Nagano Prefecture, appeared on the cover of LIFE magazine and became famous as "snow monkeys," who, come wintertime, enjoy soaking in the local hot springs.

pages 6,7: A group of monkeys soaking in the Jigokudani hot springs. The monkeys like their water hot—about 108 degrees Fahrenheit is best.

playing

Gray snow clouds move north to south above the valley known as Jigokudani, or Hell Valley. Now and then, blue sky peeps out and sunlight shines through. Immaculate snowdrifts sparkle on the southern slope of the valley. The temperature is 23 degrees Fahrenheit. The banks on each side of the Yokoyu River are in dark shadows and remain quiet, and cold, all day long. In winter, juvenile monkeys gather on the southern slope to play. They run across the snowy surface, rolling and leaping, becoming covered with snow. None of the little monkeys seem to fear the cold. Being active means being alive, and for juvenile monkeys activity is fun. One starts to play, then the others join their playmate in the frolic. They climb a tree, grab a branch with their hands, give it a shake. They swing around on it, hang from it by foot, and tumble down onto the snow-covered slope. Then they get up, run back to the tree, and climb it again, and again, and yet again as the play grows more lively. They leap from branch to branch among the trees. They work up some momentum, shake the branch, scrutinize it, then, in an instant, their bodies are flying forward, and they kick out with their feet. And without fail, they grab the next branch with their hands.

page 10: A sharp slope cuts into the valley along an active fault line. In winter, the light between sunrise and sunset is abbreviated by the steep topography and dense snowfall.

page 11: The snow clouds have disappeared. Last night's snow starts to glisten on the southern slope.

pages 12,13: Juvenile monkeys gather on the sunny southern slope. They look as if they are planning some mischief.

17

This infant monkey wants to go play, but its mother does not want to let it go.

opposite: Too small to play with the others, this infant remains content in its mother's arms.

preceding pages: Their play grows livelier as their bodies gradually warm up.

Monkey children relax in the sun's warmth.

Adults relax by grooming each other.

preceding pages:

left: In a pool of sunlight on the southern slope,
a male monkey seems concerned about his mouth.
right: A female monkey examines her nipple.

Young males play at mounting.

opposite: **A** little monkey relishes a small, tasty branch.

snowballs

The snow on a steep slope has broken away, creating large chunks of snow. The juvenile monkeys run toward and then away from the chunks as they compete with each other for possession. One stops and picks up a ball of snow; it is just the size of a small stone. He bites into the snowball, then rubs the ball on the surface of the snow and forms a bigger snowball. He stops, and looks at the ball. Then he rubs it some more, and looks again. His playmates gather around him, but he runs away with the snowball. The others quickly catch up. They become tangled and fight over the snowball. This struggle for possession is a popular pastime for juveniles, not just for snowballs but for small stones or branches as well. The snowball is pushed and rolled across the snow, carried, and stepped over, and grows bigger and bigger as it picks up more snow. While human children would have a snowball fight or make a snowman, these monkeys delight in the sheer acceleration of their play with the snowball. And when the snowball gets too big or breaks, they stop. They leave it behind and run off, once again caught up in chasing one another across the slope.

Trying to keep possession, a young monkey hangs on to his snowball.

The juvenile monkeys put every ounce of energy into their play.

following pages:

left: It's mostly the young male monkeys who join together
and plunge into their sport.

right: This monkey's snowball has grown too big to play with.

pages 34,35: Under attack by an older brother, the younger
monkey spars and plays.

Soaking up the warmth of the sun, this monkey idly basks in the sun.

opposite: An adolescent male sunbathes behind a snowdrift
that blocks the north wind.

preceding pages:

left: A quiet moment. The sun is about to set behind a ridge.
right: A juvenile monkey shakes a branch, and the snow on
it falls into the cedar woods.

following pages: Male monkeys with
blood ties: The family resemblance
is very clear.

A female savors the flavor of spring concentrated in the young leaves.

preceding pages:

left: An infant picks up a cone-shaped cedar fruit, but doesn't eat it.
right: A monkey strips off the tough cedar bark and gnaws at a trunk that has already suffered numerous chew marks. In the winter, when food is scarce, bark becomes part of the monkeys' diet.

Spring comes to the forest, and the trees begin to bud.

following pages: Wild cherry blossoms are sweet and tasty when they are no more than half open.

spring

Spring first comes to Jigokudani at the Yokoyu River. The river rushes down the slope collecting snowmelt from the backwoods of Shiga. As the water races through the valley, cascades over dams, and fills the dry riverbed, it takes on a bluish cast. Then spring plumps up the winter buds, blooms the flowers, and rekindles the pastel colors among the trees along the river. Day by day, the sunlight grows perceptibly gentler. Like smoke, the pale colors gradually cover and soften the sharp slopes of both riverbanks, then the trees—still in their winter hues—and finally the high ridges. In springtime, food for the monkeys is amply replenished: alders, willows, elms, and several varieties of maple. Each monkey climbs its favorite tree, sits on a comfortable branch, and sets about picking and eating delicious buds, leaves, and flowers. The trees in the valley cannot be harvested, so a variety of them has flourished and undergrowth covers the forest floor. Mountain vegetables sprout on the southern slope. Finding butterburs, the buds of the Japanese angelica tree, the Chinese matrimony vine, and other vegetation, the monkeys move from one mountain to another. In the spring, babies are born to the troop.

A pregnant monkey rests in the sun. The snow monkey's gestation period is approximately 173 days.

opposite: Monkeys foraging in the trees for food.

Suckle then sleep. Wake up, suckle again, then sleep. For the first month of its life, a newborn's time primarily is spent sleeping and feeding.

opposite:

A newborn looks at his first morning.

following pages:

left: For several days after its birth, the infant never leaves its mother. It clings to any and every part of her body: breast, back, bottom, head, and face.
right: Babies are born between April and June. Here, an infant monkey enjoys the comfort of its mother's arms.

left: Its eyes are beginning to see. Everything is so curious.

right: If a baby becomes uneasy, it turns away from the world and clings to its mother's breast.

newborns

A newborn suckles, clinging hand and foot to its mother's breast. The female sits, lightly holding her baby with her arm, occasionally looking at it as if to remember it. The baby cannot yet see; safe in its mother's arms, it stretches its thin and fragile limbs and bends itself over backward, dropping its red face back. It sucks milk with its blue eyes wide open. The baby doesn't free its mother's nipple easily even though it is full. It falls asleep while holding the nipple in its mouth, and moves its mouth while dreaming. The baby demands milk and the female breast-feeds: a blood relationship of this kind, based on reciprocation, is the starting point for the Japanese monkey, whose way of life is determined by a handful of matrilineal families forming a troop. Some mother-baby pairs stay close and bathe together in the spring sunshine on the Yokoyu River's banks. A week after birth, babies begin to part from their mothers and move by themselves. The growth of their legs is slower than their arms, so they toddle. First they crawl with their hands. Then they use their feet. They still cannot balance well. Their heads are heavy, too. They fall easily. They cry. The females in the troop hear their cries and come to retrieve them. But the babies still want to move; moving is being alive. So they move around, cry, are taken back, and venture out again among the large and small rocks piled up on the dry riverbed. Their downy hairs stand out and look golden and transparent in the sunlight. A child monkey from the troop notices the baby and walks by to sniff it. Babies meet each other, their faces touch, one turns around, and the other shakily follows.

If its mother feels at ease, the baby relaxes, too.

Females keep watch at all times, but they are especially cautious when they have babies.

a distraught mother

A female descends the slope carrying a baby. She walks down clumsily, clutching her baby to her chest with one hand. She stays toward the rear of the group. She descends for a while, then stops, sits down, and stares at the dry riverbed below. Then she stands up again, shifts the baby to her other arm, and breaks into a run. Dragged along the slope, bumping into rocks and trees, the baby comes along with her, but it is dead. Again the female stops. After a while, she goes back up the slope. She leaves the baby behind, stranded on a rock along the way. She returns, takes it to her breast and clumsily chases after the group descending ahead of her. On the dry riverbed and away from the troop, she eats some green grass.

The baby has been put aside. Its overall color has already changed and darkened. Its body is battered from being carried around, and only the roundness of its head yet remains. Flies gather. The female gradually moves away from the baby while she continues eating the green grass. She frequently looks back, and even returns a few times. Unceasingly, the flies land on the baby. She squats down, and, as if she is trying to catch the flies, she waves her hands and her eyes wander across the air. In the parent-child relationship between this mother and her baby, she is not receiving any reciprocation from the baby for her concern. The baby is already cold and still, but the mother seems unwilling to accept its death.

Springtime reveals a number of such mother-infant pairs. Because of her blood relationship with the newborn, the mother's instinct is to carry it with her, even though it is dead. Eventually, however, each leaves her motionless infant behind. While she trudges around from place to place, foraging through the mountains with her troop on the lookout for something good to eat, she will forget her dead infant and abandon it somewhere on a mountainside.

A female, a baby, and an older sister.

A female separates herself from her baby while eating. Right away, the older sister comes closer to the baby.

females

The matrilineal snow monkey family is a tight collection of blood-related females—including grandmothers, aunts, sisters, cousins, and nieces—around a parent-child core. The baby in a snow monkey family becomes very close to its mother first, then gradually to the blood-related females around it through frequent and repeated association. In the spring when babies are born, these blood-related infants play together. Naturally, the mothers of the playing infants become more familiar with each other, which tightens the female-dominated family. In particular, when a female who already has a daughter gives birth to another baby, the older sister spends a lot of time playing with the new family member. Older brother monkeys also play with the baby, but older sisters become more involved. If the baby cries, its older sister will embrace it, groom it, and attempt to play with it. If the baby is still crying, the older sister will bring the baby to their mother. The baby is held, suckled, and quieted down. Its sister will stand by and watch, but she cannot help because she's not old enough to nurse the baby.

The baby grows up among females in the matrilineal family and is always observing them. When the babies become children, they continue to watch the females and develop relationships with new babies, and eventually grow into adulthood. This cycle is repeated throughout the troop.

opposite: Looking for fresh greens, a mother and infant move in symmetry.

This older sister monkey likes to hold the new baby all the time.

opposite: A baby and its one-year-old sister. Their mother is eating just a small distance away.

following pages: Young snow monkeys at play. Older brothers and cousins jump in, and the play gets rough. Sometimes, the babies begin to cry.

The matrilineal family is extremely close.

opposite: These infants are about two months old. They still don't move very actively; they gather and play like babies.

preceding pages: The day is warming up, so the monkeys remain in the shade of a tree.

In the mountains, the green in the foliage deepens in darkness and intensity.

An adult monkey's skeleton lying in a valley.

Male juvenile monkeys compete and test their strength during their play.

opposite: **A** young male stands up and walks. Bouncing a
little, he puts a lot of energy into this activity.

Males mount each other and confirm each other's strength.

In summer, it is cool by the water.

males

In the entire area of Jigokudani, there are three monkey troops. A troop's territory extends some six to twelve square miles through the mountains. A troop moves through the territory together, with the matrilineal families almost at its center. Adult males keep to the periphery. They move independently, or in the company of child and juvenile monkeys, or with another adult leading a group of younger monkeys. So the shape of the group takes on a variety of forms. As females associate with each other, males associate with each other within the peripheral male community of the group. The territories of the three troops sometimes overlap, so the troops associate with each other as well. Unlike the females, the males have gathered and played together since they were babies. Play is repetitive and often gets rough. They move among the territories, play, grow up, become adults, and mutually affirm their lives within the troop.

opposite: Suddenly, an adult male bites a young one.

After a rain shower, a baby licks water droplets from
the tips of its mother's hair.

The river swells after a typhoon.

Even in the summer the water of the Yokoyu River is chilly.

Just a brief stay in the water is enough.

A female with a newborn experiences a delay in the change from winter hair to summer hair.

opposite: A baby catches a bee, and immediately throws it away.

Making sure where it is to land, a monkey takes a big leap. When a monkey gets to be about a year old, it can do this without difficulty.

opposite: **A** baby who was playing at the waterside has been retrieved by its mother.

The downy hair of winter is replaced with summer hair.

opposite: **A** baby male mounts a female.

A female baby.

The summer ends, and the winds blow more strongly.

Autumn is the monkeys' mating season. The faces and bottoms
of adults, both male and female, turn bright red.

opposite: A male vigorously jostles a tree.

species

A male and a female stay affectionately close along the periphery of the group. The female is with a baby and a two-year-old child. In the autumn mating season, males seek a mating partner within the troop. They swagger around inside the group, ostentatiously displaying their maleness to the females. Some males have been born within the troop; others have joined from other troops. The male is certain to find a female in the troop if he is independent and suitably virile. Regardless of whether the female is carrying a baby or a child, regardless of who the father is, regardless of how many relationships the male has had—none of this matters to him. If the female selects him as her companion, they pair off, stay together around the edge of the group, and repeatedly mate. They become more affectionate toward each other while grooming between matings. If the female has any babies or children, the little ones observe the couple off to the side. Soon fall will end, and winter will arrive. Then, once again, spring will come around and the babies will be born.

While mating, this female holds a baby. The male keeps watch around them.

opposite: **A** frustrated female jumps on the back of a male.

The male on the left grooms the female on the right. He picks out the lice eggs and eats them.

He parts her hair and finds a lice egg. He picks it out and chews it.
He has keen vision, and his fingers move skillfully.

Adults grooming each other.

A male child watches two adults mating.

A mother and child carefully watch a nearby male.

The bright red posterior of a male snow monkey.

One young female carefully grooms another.

Fall is coming to Jigokudani.

Young males relaxing together.

An adult male running through the mountains.

In the fall, the mountains are full of food.

opposite: Eating the berries of a sawashiba. There are plenty of them.

following pages:

left: A young monkey eating a twig that it found in the dry riverbed.
right: Her cheek sack filled with acorns, a female sits down and then eats each acorn slowly. She will split the shell in two with her teeth and spit it away.

Fruit still remains at the tip of this dogwood branch. Cautiously, a child monkey extends its hand to reach it.

preceding pages: A lone male having a quiet moment away from the troop.

Snow monkeys eat mushrooms, too.

following pages: Autumn is at its peak, and the mating season is at its height.
This couple stays affectionately close.

A male taking a brief rest.

Adult males approach and the young males move aside to let them pass.

A male sniffs a female's bottom.

The male on the far left bides his time while others are groomed.

The male uses the female as a takeoff point and leaps over part
of the Yokoyu River. The female stands stock-still with fear.

A mother is grooming her boy monkey. The baby joins in and does likewise.

following pages:

left: The male is larger, heavier, and stronger than the female.
right: A baby has been bitten by a male. Many monkeys get
injured during autumn.

A male child monkey is bleeding.

opposite:
The male licks the blood of the child monkey off of a leaf.

As soon as the females' estrus ends, the mating period ends.

opposite: If the female is not sexually excited, the male cannot mate with her.

preceding pages: From the ridge, a young male watches the troop down on the dry riverbed.

winter comes

A northwest wind is blowing. Thin stripes of clouds whirl in the sky high overhead. The movement of the gray clouds is swift. They spread over the ridge and move toward the backwoods of Shiga. The valley wind roughens. The wind sprays the dry yellow, red, vermilion, brown, and purple leaves, sweeps them together, and hurls them against the slope. The monkeys' winter coats are stirred and tossed about. Then the outline of the sun is blurred. For a time, the first snow falls at Jigokudani, turning the ground white. One after another, the tumbling snowflakes cling to the long hairs. Up on the whitened slope, the monkeys go on with their eating. There is no playing together today. Hurried by the lightly piled first snow and slight chill, the monkeys in the troop keep moving, eating the dead colors that are autumn's mementos. The northwest wind howls again. The striped clouds are moving in ranks. The snow slices down.

Then suddenly it goes away, and the sun peeps out from a crack in the clouds. The snowflakes atop the monkeys' coats melt and glimmer as droplets. The babies are soaking wet. The northwest wind blows harder. Snow. For the snow monkey troops of Jigokudani, winter is fast approaching.

As winter nears, the snow monkeys look for acorns.

opposite: Eating after the first snowfall, the
monkeys devour whatever is at hand.

In winter, not much food is available, so this monkey is eating a konara, a
small oak tree, branch.

The dry riverbed has become slippery with snow. A young monkey jumps from one rock to another.

following pages: Thick gray snow clouds don't deter two
monkeys from eating the winter buds of the maple tree.

A baby getting through the winter, all puffed up in its winter coat.

The monkeys move along the slope as a group while eating. Snow makes
for an earlier sunset, and the deep blue darkness approaches.

following pages:

left: The snow slices in. This baby is crying for its mother.

right: A baby's first attempt to touch an icicle. Then it tries to lick it.

This female was born in 1963. She made the reputation of the Jigokudani Onsen
(Hell Valley Hot Springs) when she appeared on the cover of LIFE magazine.

preceding pages: From late January into February, the snow comes down hard.

Blood rushes to their heads at the hot springs.

following pages: A snapshot of a matrilineal family.

afterword

We have been associated with the Japanese monkeys of Jigokudani for years. Particularly memorable for us is the time the three of us—myself, my wife, and our daughter, who was just three at the time—paid a visit to Jigokudani. We had quite a shock when our daughter was bitten on her right shoulder by a female monkey. Fortunately, it was just a scratch. Our daughter was, of course, the smallest member of our family, which meant that her eye level was the lowest. That might have made her look more like a monkey, and the monkey might have thought our daughter was one of her troop. Or perhaps the monkey felt defensive confronting this obviously non-monkey human whose eye level was the same as its own. Nevertheless, a person should try to observe a monkey at the monkey's eye level or even lower, and that goes for any other wild animal as well. Now our daughter is nineteen years old, and she doesn't have any teeth marks from the monkey on her right shoulder. Her experience was an important lesson as we formed our attitude as humans who observe wild monkeys.

The Japanese monkeys of the Jigokudani Hot Springs are famous both nationally and internationally for their fascinating resemblance to humans. For example, the "snow monkeys" bathing in the winter hot springs wear a comfortable expression and have mannerisms that are humanlike. People tend to extend this likeness to other commonalities and feel that we completely understand the monkeys. But this is not correct. Monkeys can only be monkeys, and people can only be people. We are both primates, but if our initial point of understanding is based on human categories, then the monkeys can only become alienated from us.

We wanted to observe these monkeys as they moved about and lived their lives. If possible, we wanted to gain even a small understanding of their lives. We concentrated on gathering data on groups of Japanese monkeys within Jigokudani over a three-year period starting in 1994. *Snow Monkeys* includes photographs taken during that period while we simultaneously shot a video. The incident with our toddler furnished a clue as to how we should approach our observation of the monkeys' movements. We lowered our eye level as near as possible to that of our three-year-old daughter's, thus making our human perspective as monkeylike as we could.

We would like to express our appreciation to the persons below who provided their cooperation during the compilation of *Snow Monkeys*:

Messrs. Sogo Hara, Eiji Tokida, Haruo Takefushi, Shigenori Nishizawa, and Toshio Ogiwara of Jigokudani Wild Monkey Park; Messrs. Kenichi Mizuno, Tatsuo Kozaki, and auxiliary cameraman Keiji Hasegawa from NHK and NHK Enterprise 21; as well as the staff members of Jigokudani Korakukan Inn. We would also like to express our appreciation to Mr. Masaharu Yokoyama, manager of Shinchosha's publishing department, and Ms. Mina Karashima of Shinchosha's publishing department, as well as Mr. Norihiro Shiozawa from the editing department of *Shinra* magazine, and Mr. Keisuke Konishi, the designer, and Ms. Midori Kita. And most of all, we would like to thank the amiable Japanese monkeys of Jigokudani, who provided us with so many warm memories.

—*Mitsuaki and Hideko Iwago*

right: MINI-ENCYCLOPEDIA OF JAPANESE MONKEYS
The following mini-encyclopedia was created by Shinchosha Company in cooperation with the Kyoto University Primate Research Institute. Editorial supervision was provided by Professor Tetsuro Matsuzawa, Dr. Mitsuru Aimi, and Dr. Juichi Yamagiwa of the Kyoto University Primate Research Institute. Photographs by Mitsuaki Iwago.

Anthropoids

Old World Monkeys

New World Monkeys

Prosimians

White-handed Gibbon (lesser anthropoid)

Orangutan

Lowland Gorilla

Chimpanzee

Human

Abyssinian Colobus

Mandrill

Anubis Baboon

Vervet Monkey

Crab-eating Macaque

Japanese (Macaque)

Golden Lion Tamarin

Uakari

Squirrel Monkey

Black Spider Monkey

Geoffroy's Spider Monkey

Slender Loris

Bushbaby

Ruffed Lemur

Ring-tailed Lemur

Indri

Verreaux's Sifaka

Today about two hundred species of primates inhabit the planet. We can divide them into five large groups, including one that has human beings as its sole member.

1) Prosimians *(Suborder Prosimii)*
   Lemur, Loris, and Galago, which inhabit Asia, Africa, and Madagascar and show ancestral features unlike those of the monkey.

2) New World monkeys *(Suborder Anthropoidea, Infraorder Platyrrhini)*
   Squirrel Monkey, Spider Monkey, Lion Tamarin, Marmoset, which inhabit Central and South America.

3) Old World monkeys *(Suborder Anthropoidea, Infraorder Catarrhini, Family Cercopithecidae)*
   Japanese Macaque, Rhesus Macaque, Mandrill, Baboon, Proboscis Monkey, and Golden Snub-nosed Monkey, which inhabit Asia and Africa.

4) Apes *(Suborder Anthropoidea, Infraorder Catarrhini, Superfamily Hominoidea)*
   Divided into two families: the first is the Lesser Apes, the Gibbon family of small, long-armed anthropoids. The second is the Great Apes, the Pongid Family of large anthropoids, which has four species: Gorilla, Chimpanzee, Bonobo (Pigmy Chimpanzee), and Orangutan.

5) Humans *(Suborder Anthropoidea, Infraorder Catarrhini, Superfamily Hominoidea)*
   Only one family Hominidae, genus Homo, and species Sapiens.

Groups 2, 3, 4, and 5 are referred to as anthropoid as compared to Group 1, which is called prosimian. Platyrrhine animals exhibit a wide distance between the nostrils, which are wide open to the side, while catarrhine animals have relatively narrow-spaced nostrils. Apes are distinct from monkeys. "Hominoid" refers to the superfamily that combines gibbons, great apes, and humans. (The lesser apes are sometimes included in this superfamily.)

About 3.5 billion years ago, the earliest life is said to have been born on this planet. Originating from this seed is the great diversity and the tremendous number of life-forms, some of which have survived and some of which have come and gone.

The ancestors of modern primates made their appearance toward the end of the Cretaceous period, some 70 million years ago. Scientists believe they began to evolve from the insectivores. Hurrying along the evolutionary stream, about 50 million years ago there was a bifurcation between anthropoidea and prosimian. Prosimian is thought even today to be closest in appearance to the common ancestor of the primates. About 35 million years ago, the anthropoidea divided into platyrrhines and catarrhines. (As mentioned before, the Japanese monkey belongs to the Old World monkey group under the catarrhine infraorder.) About 28 million years ago, a division split the catarrhine into two groups; monkeys, like the Japanese monkey (an Old World monkey), and hominoids. Hominoids further divided into lesser apes, orangutans, and gorillas. About five million years ago, humans and chimpanzees broke off. Erect posture with bipedal walking (the body trunk stands straight and walking occurs on two hind legs) is a distinguishing characteristic

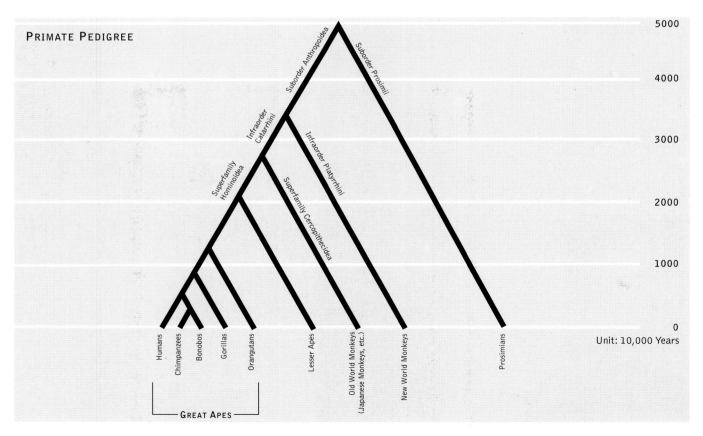

PRIMATE PEDIGREE

Suborder Anthropoidea

Suborder Prosimii

Infraorder Catarrhini

Infraorder Platyrrhini

Superfamily Hominoidea

Superfamily Cercopithecidea

5000

4000

3000

2000

1000

0

Unit: 10,000 Years

Humans

Chimpanzees

Bonobos

Gorillas

Orangutans

Lesser Apes

Old World Monkeys (Japanese Monkeys, etc.)

New World Monkeys

Prosimians

GREAT APES

separating humans from the primates. According to recent research, the fossil of an ape-man (Ardipithecus ramidus) was excavated from a geologic layer some 4.4 million years old in Ethiopia. He is thought to have stood straight and walked on two legs, so we find that even that long ago a human ancestor existed. It is also well known that human and chimpanzee DNA differ in their base configurations by only about 1.7%.

Where did the Japanese monkey come from? Old World monkeys (infraorder catarrhini, superfamily cercopithecidae, family cercopithecidae) are divided into two large groups: the cercopithecus subfamily and the colobus subfamily. The Japanese monkey (Macaca fuscata) belongs to the genus macaca of the cercopithecus subfamily. There are nineteen species of monkeys in genus macaca, including the Rhesus macaque, the crab-eating macaque, and the Taiwan macaque. Except for North Africa's Barbary macaque, all these monkeys live in Asia. And except for human beings, they are the most widely disseminated primates on

earth. Among the macaques, the Japanese monkey is famous for being the northernmost-dwelling monkey—again except for humans. They are truly hyperborean monkeys.

So, when did the Japanese monkey begin to live on the islands of Japan? The oldest fossil of a monkey found in Japan to date is about 500,000 years old, and was excavated in Akiyoshidai, Mine-gun, Yamaguchi Prefecture. So far, in addition to this find, Japanese monkey fossils have been discovered in only a few places: Kisarazu City in Chiba Prefecture, Kokawa-cho in Ehime Prefecture, Yugi-cho in Hiroshima Prefecture, Hachiman-cho in Gifu Prefecture, Fujisawa City in Kanagawa Prefecture, Katsuo-cho in Tochigi Prefecture, and Shiriyazaki in Aomori Prefecture. This means that the ancestor of the Japanese monkey existed in Japan at the latest during that time frame. But at that time the main island of Japan shared a landmass with the Korean peninsula. Moreover, a two-million-year-old fossil of the genus macaca has been found in neighboring China, so it is possible that the Japanese monkey lived in Japan even earlier than the fossil record indicates.

The Japanese monkey's ancestor is said to have come to Japan across the continent from Southeast Asia over a very long period of time.

DISTRIBUTION OF JAPANESE MONKEYS

The Japanese monkey is widely distributed mainly throughout the broadleaf forests of the Japanese archipelago except for Hokkaido and the Ryukyu Islands. They are not found in Hokkaido, probably because it is a very cold region and is separated from the main island by a narrow strait. The sunny Ryukyus, on the other hand, are separated from Japan by the deep sea.

How many monkeys are in Japan? There is no accurate data on the population count.

On Yaku Island, a subspecies called the Yakushima monkey lives. They are a little smaller than the Japanese monkey, and their hair color is darker.

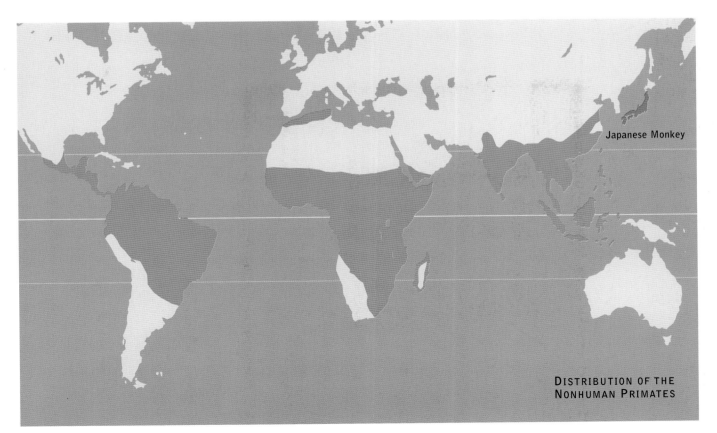

Japanese Monkey

DISTRIBUTION OF THE
NONHUMAN PRIMATES

CHARACTERISTICS OF THE JAPANESE MONKEY

A characteristic common to all primates is the ability to grasp an object with all four limbs. A common characteristic of the cercopithecidae (Chercopithecus) family is its calloused rump. A common characteristic of the cercopithecidae subfamily is the cheek sack, which the monkey can pack with food, then move to a safer place for chewing.

The average weight of males is 26 to 33 pounds (12 to 15 kilograms) and for females 18 to 29 pounds (8 to 13 kilograms). Height (head and trunk) ranges from 22 to 24 inches (54 to 61 centimeters) for a male and from 19 to 24 inches (47 to 60 centimeters) for a female. The tail is short, at 3 to 5 inches (8 to 12 centimeters) for a male and 3 to 4 inches (7 to 10 centimeters) for a female.

Teeth are the same as ours; 20 deciduous and 36 permanent. Body temperature is 38.6 degrees Celsius (101.4 degrees Fahrenheit), slightly higher than ours. Also characteristic of the Japanese monkey is a reddening of their faces and bottoms as they become adults. In particular, this rubescence deepens during the fall-winter period of sexual activity. This skin phenomenon is not due to pigmentation, but is the result of capillary dilation induced by sex hormones, causing the blood under the skin to become more visible.

Their body hair is two-layered. Even if the top layer gets wet, the softer inner hair is protected and chill-proof. Some Japanese monkeys bathe in hot springs. The reason they are not chilled after bathing is because of this coat structure.

Primate eyes line up almost horizontally right to left, so the Japanese monkey can see objects three-dimensionally although their visual field is narrow. Their vision is very similar to human vision. Their color sense is also similar; they can distinguish between a variety of colors.

As for hearing, human beings can sense air vibrations and sounds within a frequency range of between 20 and 20,000 hertz. Humans cannot hear the higher frequencies above 20,000 hertz. The Japanese monkey, however, can hear high frequency sounds up to about 50,000 hertz. Very little is known about their sense of smell.

THE DIET OF THE JAPANESE MONKEY

The Japanese monkey eats a variety of foods, but mainly it consumes plants. In spring, they prefer to eat new buds, young leaves, and flowers; and in the fall, fruit and seeds. Insects are favorites, too. Japanese monkeys living in snow country survive the winter by eating bamboo leaves, winter buds, and bark. But these do not supply sufficient nutrition, so the monkeys apply a method of mass consumption throughout the fall to store up fat; they supplement their nutritional requirements by gradually dissolving this fat. If during the fall the fruit from the trees is in short supply or if the springtime snowmelt is delayed, they are in danger of starving to death. The monkeys do not seem to store any food.

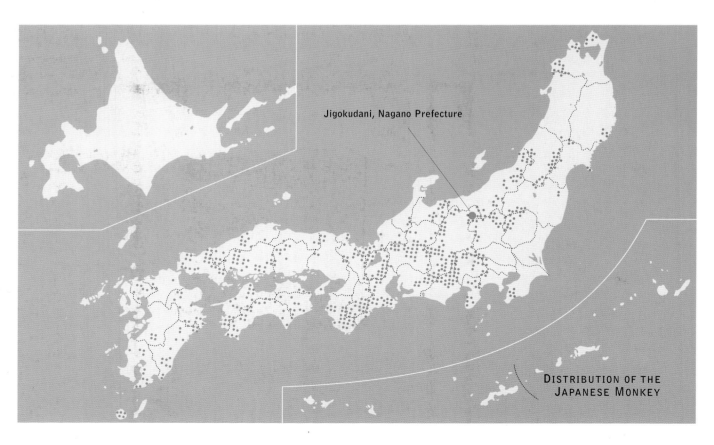

Jigokudani, Nagano Prefecture

DISTRIBUTION OF THE JAPANESE MONKEY

A Japanese Monkey's Day

The Japanese monkey is active during the daytime. At night, they sleep in the trees by embracing one another or, if one is alone, by squatting and holding its own knees. They don't make nests, and every day they move to a new place to spend the night. For overnighting during snow season, they select such places as the branches of an evergreen tree that's perched on a slope protected from the wind.

From dawn to the dusk, they slowly move about as a group while searching for food, taking rests, or playing. This group movement is termed "ranging," and a particular place used for this activity is called a "home range."

The breeding season for Japanese monkeys is from October to February. A male mounts a female, and for five seconds or so at a time for about a dozen minutes he engages in vigorous activity until he ejaculates. Childbearing extends from April to July. A female gives birth for the first time between four and six years of age. Typically, in the wild, these monkeys give birth about once every three years. A pregnancy lasts on average 173 days and usually results in one baby. It is rare to chance upon a wild Japanese monkey giving birth. A monkey gives birth alone without help as the newborn passes down the birth canal. She stretches her legs, pulls the baby out with her hands, and holds it up to her chest. The newborn has hair all over its body and its eyes are wide open. After a few days, the infant starts following moving objects with its eyes and tries to crawl. After two weeks, it begins toddling, and after a month it starts to swallow food other than its mother's milk and to occasionally play away from its mother. After three months, it begins active play, though sometimes it returns to its mother to suckle. Actual weaning occurs about the sixth month. Nevertheless, most babies continue suckling for about a year, and a small amount of mother's milk seems to be available. The death rate is at its highest during the youngster's first year.

Males start their breeding activity when they are six years old or even older, which is later than for the females. A strong male that remains with a troop can maintain its ability to ejaculate even after the age of twenty-five. Females usually give birth for the last time when they are in their early twenties.

The snow monkey's life span is typically twenty-five years, or at most thirty years. Enfeeblement interferes with their ability to keep up with the group, and they seem to quietly pass away.

Social Life of the Japanese Monkey

Japanese monkeys live in a group called a troop. A troop can consist of from ten to more than a hundred monkeys. In Japan's snow country, a troop's maximum size seems to be about ninety. An average-size troop has a few adult males, two to three times as many adult females, and the juveniles. This is called a multiple male-and-female group. Females spend their entire lives in the troop where they were born. Many males, however, leave their birth troop when they reach adulthood. Some join male-only troops, and some live independently, but they later join a troop. Throughout their lives, they seem to move between several troops. Of course, there are exceptions.

A troop's actual uniformity depends on the kin relationships between the adult females and their children. Theirs is a matrilineal society. The social positions for males are determined by their ages, physiques, and personalities, including their temperaments, but for females, kin relationships affirm their social positions.

Japanese monkeys communicate with each other by their grooming behavior. They groom all the time; this includes self-grooming. Grooming serves not only a sanitary function, but is also a means of comforting others and strengthening or repairing relationships between individuals. Incidentally, we often talk about a monkey's flea picking, but they are actually picking lice eggs and eating them throughout the grooming session.

Monkeys also communicate by other means, such as facial expressions, voices, and gestures. For example, when a monkey narrows its forehead by raising its eyebrows and stares with a half-open mouth without showing its teeth, that expression signifies a threat. Sometimes sounds like "goh, goh" are vocalized, too. On the other hand, their expression for calming another monkey down is the so-called "lip smack," made by rapidly opening and closing the mouth again and again.

The Future of the Japanese Monkey

In recent years monkeys have caused some problems. They have begun appearing near houses and have damaged agricultural products. Researchers feel that because the monkeys' forest habitats have dwindled due to land development, they have been forced to live near the surrounding villages. In addition, there have been cases where monkeys have grown used to people because people feed them.

In Attachment II of the CITES, which governs the international trade in endangered animals and plants, the Japanese monkey is listed. A subspecies, the Yakushima monkey, is designated as an endangered subspecies in the Red Data Book of the International Nature Preservation League. In six locations within Japan, Japanese monkeys are designated as a natural monument. They are a protected species, but they can be caught and killed as nuisance animals if the prefectural government permits. Today, five thousand or more monkeys are either caught or killed each year because of the damage they inflict. This number may exceed that of the population's natural growth. In addition, because their natural habitats are closer to human residences, the monkeys' risk of infection from human diseases grows more likely.